Camilla Albertina Dantas de Lima

Osteoporosis, the Immune System and Genetics

Camilla Albertina Dantas de Lima

Osteoporosis, the Immune System and Genetics

A new approach to an old acquaintance

ScienciaScripts

Imprint
Any brand names and product names mentioned in this book are subject to trademark, brand or patent protection and are trademarks or registered trademarks of their respective holders. The use of brand names, product names, common names, trade names, product descriptions etc. even without a particular marking in this work is in no way to be construed to mean that such names may be regarded as unrestricted in respect of trademark and brand protection legislation and could thus be used by anyone.

Cover image: www.ingimage.com

This book is a translation from the original published under ISBN 978-620-2-17587-6.

Publisher:
Sciencia Scripts
is a trademark of
Dodo Books Indian Ocean Ltd. and OmniScriptum S.R.L publishing group

120 High Road, East Finchley, London, N2 9ED, United Kingdom
Str. Armeneasca 28/1, office 1, Chisinau MD-2012, Republic of Moldova, Europe
Printed at: see last page
ISBN: 978-620-7-34172-6

Table of contents:

"Whether this DNA records stories about music, sports or Machiavellian microbes, the narratives together tell a more intricate story about the emergence of human beings on Earth: why we are one of nature's most absurd creatures, as well as its greatest glory." Sam Kean

Osteoporosis, the Immune System and Genetics: A new approach to an old acquaintance.

ISBN: 978-620-2-17587-6

New Academic Editions

Author

Camilla Albertina Dantas de Lima - PhD in Genetics from the Federal University of Pernambuco, MSc in Cellular and Molecular Biology and Specialist in Clinical Pathology from the University of Pernambuco and Graduated in Biomedicine from the Federal University of Pernambuco. She works mainly in the areas of Human Genetics and Molecular Biology with an emphasis on Rheumatology.

Co-authors

Paula Sandrin-Garcia - PhD in Biological Sciences (Genetics) from the University of São Paulo, MSc and BSc in Biological Sciences from the State University of São Paulo. She works as an Adjunct Professor II in the Department of Genetics at the Federal University of Pernambuco and is a permanent member of the Postgraduate Program in Genetics and the Postgraduate Program in Biology Applied to Health, both at the Federal University of Pernambuco. His research focuses mainly on autoimmune diseases such as lupus, arthritis and osteoporosis.

Alexandre Domingues Barbosa - PhD and MSc in Biology Applied to Health from the Federal University of Pernambuco and a degree in Medicine from the Federal University of Pernambuco. He works as a Rheumatologist and coordinator of the Osteoporosis and Osteometabolic Diseases outpatient clinic at the Rheumatology Department of the Hospital das Clínicas of the Federal University of Pernambuco.

Maria Aparecida Barreto Lopes Seabra - PhD in Pharmaceutical Sciences from the Federal University of Pernambuco, MSc in Biochemistry from Kansas State University and BA in Chemistry from the Catholic University of Pernambuco. She has experience in Chemistry and Cell Biology, with an emphasis on models for the study of biomarkers using cell culture.

Dedication

To the patients who, in the midst of their physical pain, give of themselves in the hope of a cure for all.

To Caio for his sweet and constant companionship throughout this project.

To Welton for all the love and encouragement that has made our days so happy.

Summary

Osteoporosis (OP) is an osteometabolic and multifactorial disease related to genetic factors in more than half of the diagnosed cases. Recent genome-wide association studies have shown the close relationship between bone homeostasis and the immune system, confirming the influence of various cytokines and their pathways related to the characteristic phenotypes of the disease. In order to functionally validate these findings and clarify the participation of these genes in the development of postmenopausal OP, this study investigated the relationship between single nucleotide polymorphism (SNP) in the IFNG gene and susceptibility to the disease. *In* addition, *in vitro* tests were carried out on human osteocarcoma cells (SaOs-2) to study their viability profile and calcification processes. Our results showed the SNP IFNG -1616 (G>A) (rs2069705) associated with susceptibility to OP and the cytokine IFN-Y increasing the calcification process in SaOs-2. Following these analyses, we can demonstrate the importance of the SNP studied in OP susceptibility, as well as helping to clarify the role of IFN-Y IN the calcification process in bone cells.

Keywords: SNPs, Osteoimmunology, immune system, IFN-y.

Foreword

As they grow, scientific advances bring with them not only answers to previously raised doubts and hypotheses, but also, and above all, relief for many of humanity's pains. In the field of health, they change certainties, add a lot of questions, answer some of them, altering, remodeling, constructing and deconstructing concepts, changing perspectives, opinions and treatments. Such is science. And within it, Genetics, which since the beginning of its studies with Mendel and Miescher has overcome the barrier of being discredited to becoming a latent answer to so many desires in the biological and health sciences. Like a matrioska, it reveals an unknown world to be unraveled, studied and analyzed. And repressing the initial enchantment a little, we continue as researchers, in a magical world, but mainly of proofs, counter-proofs, theses and antitheses to be investigated with magnifying glasses, or rather, with high-resolution microscopes. So that we can experience all this meticulously orchestrated, designed and maintained organization, telling us about pathophysiology, clinical characteristics, treatments and cures. This is also the case with Genetics, the Immune System and Rheumatology. Enjoy!

Camilla Lima

Chapter 1

Introduction

Osteoporosis (OP) is a degenerative, multifactorial osteometabolic disease characterized by a reduction in bone mineral density (BMD) and deterioration of bone microarchitecture. The main recurring phenotypes are increased skeletal fragility and consequent susceptibility to fractures. Of the various systemic and local factors involved in the etiopathogenesis of the disease, the association between hormonal components, genetic *background* and the immune system has been shown to be the main factor in maintaining bone tissue homeostasis. However, despite the recognition of the importance of these factors and the knowledge generated from *genome-wide* association studies (GWAS), the results regarding the action of many of these components are still contradictory. In addition, few functional studies have been carried out to better understand the processes involved in the establishment of the disease.

Since the term Osteoimmunology was first used by Arron and Choi (2000), several studies have linked immune system proteins to the balance between the activities of osteoclasts (bone resorption cells) and osteoblasts (bone replacement cells). Pro- and anti-inflammatory cytokines have been associated with bone homeostasis, acting to induce or inhibit tissue resorption by controlling the proliferation of these cells. Alterations in the genes of these cytokines, such as single nucleotide *polymorphisms* (SNPs), can alter both the structure and functionality of these proteins and consequently modify the biological mechanisms in which they are involved.

The importance of understanding the mechanisms of action of these proteins lies not only in gaining a new insight into the pathogenesis of OP, but also in the possibility of their clinical applications. Certain cytokines can serve as biomarkers in the monitoring and prognosis of the disease, and may even become therapeutic targets for its treatment, given the side effects caused by current therapies. Some pro-inflammatory cytokines, such as the receptor ligand for nuclear factor *kappa* B

(RANKL), have already been reported as potential targets for diagnosis and therapy in the fight against reduced BMD and the occurrence of fractures. The aim of this study was to investigate whether the IFNG -1616 (G>A) SNP (rs2069705) in the IFN-Y pro-inflammatory cytokine gene is related to altered levels and susceptibility to primary postmenopausal OP. In addition, we verified the influence of this cytokine on calcification processes and the viability of cells of the osteoblastic lineage.

The polymorphism studied showed susceptibility to OP and differences in protein expression levels were also detected between the groups of patients and healthy women. In addition, the cytokine IFN-Y influenced the calcification process and viability in human osteosarcoma cells (SaOs-2). In this way, we demonstrate the importance of the SNP studied in susceptibility to OP, as well as helping to clarify the role of IFN-Y in the calcification process of cells of the osteoblastic lineage.

Chapter 2

Literature Review

2.1 Characterization of Osteoporosis

According to the World Health Organization (WHO), OP is defined as a systemic disease of the skeleton, characterized by a decrease in bone mineral density (BMD) and deterioration of the tissue's microarchitecture. This leads to increased bone fragility and greater predisposition to fractures (Kanis, 2008; Zaia, 2015; Rocha-Braz and Ferraz-de-Souza, 2016) (Figure 1). Although prognosis, diagnosis and monitoring techniques are closely related to BMD and estrogen and calcemia levels, it is known that many factors are involved in the pathogenesis of the disease, classified as clinical factors of bone and non-osseous origin (Aspray, 2014).

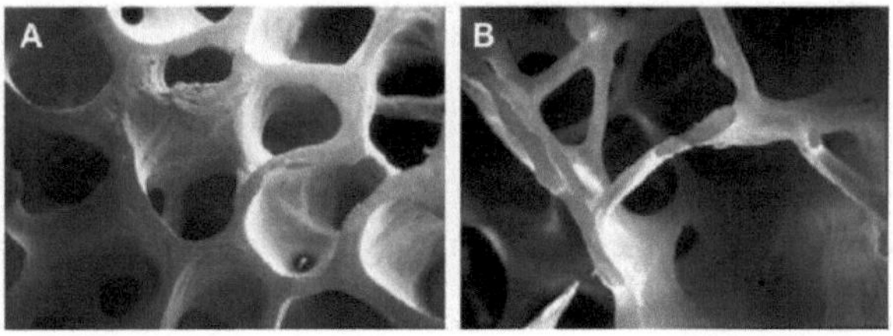

Figure 1: Micrograph of bone. A: Normal bone. B: Osteoporotic bone. Source: Golob & Laya, 2015.

Elements such as advanced age, Caucasian or Asian ethnicity, low body weight, family history of the disease, smoking, excessive alcohol consumption, sedentary lifestyle, low vitamin D levels and insufficient calcium intake make up the main primary risk factors related to the development of the disease (Golob and Laya, 2015; Greenblatt et al., 2016). OP does not always begin with a direct change in the composition and arrangement of bone tissue, as in primary postmenopausal OP. In some cases it can be secondary to other pathologies such as hypogonadal syndromes, thyroid alterations, diabetes mellitus, rheumatoid arthritis, malabsorption syndromes, chronic liver, kidney and lung diseases, hematological disorders, as well as excessive consumption of medications, especially glucocorticoids (Pereira, 2009; Golob and Laya, 2015).

In addition to these clinical factors, components of other biological systems have been shown to be relevant in recent research related to skeletal disorders, such as the genetic profile of patients and the behavior of the immune system and its proteins (Arron and Choi, 2000; Zupan et al., 2013). With whole genome mapping, several immune system genes have been associated with decreased BMD and increased bone fragility, presenting possible new prognostic and therapeutic markers for the disease (Liu et al., 2014; Clark and Duncan, 2015; Qin et al., 2016).

The importance of deepening these basic studies is even more evident given the underdiagnosis and undertreatment of OP, which aggravates this public health problem (Roush, 2011; Golob and Laya, 2015). Within the group of neglected diseases, primary postmenopausal OP is particularly worrying due to its high rates of involvement and significant decrease in the quality of life of older women, which may even lead to death, in addition to the symptoms of severe pain and increased morbidity (Iolascon et al., 2013; Baccaro et al., 2015). Another factor to consider is the heterogeneity of the methodology used for diagnosis, which contributes to the inaccuracy of the disease's prevalence figures (Golob and Laya, 2015). Based on some population studies carried out in Brazil and the United States, the average rate of women affected after the age of 50 varies between 15-35% (Pereira, 2009; Golob and Laya, 2015). According to the WHO, by the year 2050 the number of hip fractures is expected to increase by 400% in the population aged up to 60 years, and by 700% in the population aged over 65 years, for Latin America alone (Cunha et al. 2011).

As this pathology has multifactorial causes and is often asymptomatic, its diagnosis can be inaccurate and difficult to predict. Initially, this diagnosis was based solely on BMD indices obtained through dual-emission X-ray absorptiometry (DEXA), with the result later classified according to the number of standard deviations above or below the patient in relation to the average BMD of a reference healthy young population, the T-score (Golob and Laya, 2015; Zaia, 2015). T-scores above -1.0 are considered normal, between -1.0 and -1.99 as osteopenia, between -2.0 and 2.49 as advanced osteopenia, below -2.5 as OP and T-scores below -2.5 associated with one or more fractures are classified as severe OP (Figure 2) (Baccaro et al., 2015; Golob and

Laya, 2015).

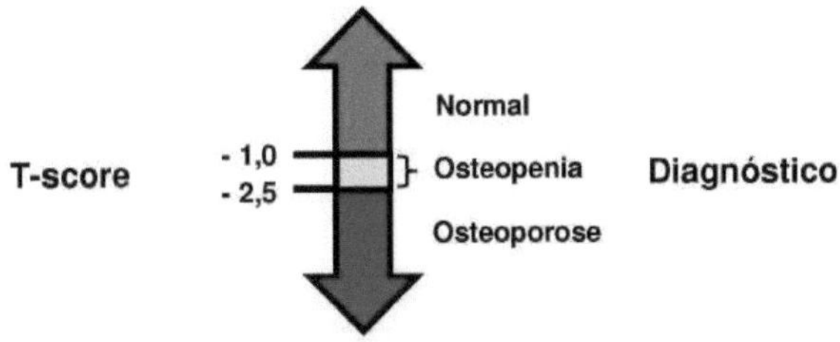

Figure 2: T-scores resulting from DEXA scans and their respective associated diagnoses according to the WHO.

However, other studies have shown that BMD analyzed in isolation does not have high diagnostic efficacy, given that not only the amount of bone tissue is important in the homeostasis of the system, but also its microarchitecture, which is responsible for correcting bone fragility due to resorption of the tissue matrix (Aspray, 2014; Zaia, 2015).

Other components that can help in this context are biochemical markers of bone *turnover* (Greenblatt et al., 2016). The variation in these proteins resulting from bone cell activity can be detected more quickly than changes in bone mass (Chesnut

14 et al., 1995; Greenblatt et al., 2016). The main markers used in OP clinical practice and research are alkaline phosphatase (ALP), osteocalcin (OC), tartrate-resistant acid phosphatase (TRAP) and type I collagen carboxyterminal telopeptide (CDX) (Seibel, 2005; Zhou et al., 2014; Zhou et al., 2015). Additionally, levels of parathyroid hormone (PTH), 25(OH) hydroxyvitamin D and calcium are monitored in this process (Zhou et al., 2014; Zhou et al., 2015). However, these elements are generally restricted to monitoring and not to the diagnosis or prognosis of the disease (Greenblatt et al., 2016).

2.2 Epidemiology of Osteoporosis

According to the latest IBGE census (2010), Brazil has around 51 million people over the age of 55, 10 million of whom have OP. In 2010 alone, 74,000 patients were admitted to the public health system due to fractures resulting from the disease. Brazilian hospital spending on fracture patients costs around 24,000 reais per patient and government spending on the disease amounted to 80 million reais in 2010 alone. According to the *National Osteoporosis* Foundation (NOF), the approximate annual expenditure for 2015 in the United States was 20 billion dollars (Roush, 2011; MS, 2014). Despite representing a serious public health problem in Brazil and around the world, OP is considered a neglected disease, given that in Brazil only 24% of diagnosed patients receive specific therapy, and around 20% die within a year of fracture due to diseases secondary to falls and morbidity (Iolascon et al., 2013; MS, 2014).

In this context, accurate diagnosis and specific treatment are extremely important, given that adherence to the treatment suggested by the clinician can only be assessed after a minimum of 6 months, which is sufficient time for the risk of falls and subsequent fractures (MS, 2014). In addition to lifestyle changes such as exercise, calcium intake in the diet and a reduction in alcohol and cigarette consumption, drug therapy is indicated mainly for post-menopausal women with BMD values below -2.5 T-score; post-menopausal women with BMD between -1 and -2.5 with a predicted risk of hip fracture of 3% or 25% in any region over the next 10 years and post-menopausal women with a history of fractures (Roush, 2011; Golob and Laya, 2015).

2.2.1 Epidemiology of Osteoporosis Fractures

OP has become a major public health problem mainly due to bone fragility fractures and their association with increased mortality and morbidity, which result in a considerable expense for the global economy (Kanis et al. 2011; Vijayakumar and Dietrich 2016). Osteoporotic fractures occur especially in older individuals, and as life expectancy is increasing worldwide, the occurrence of this type of fracture is also likely to rise. It is estimated that the number of people over 65 will double by the year 2050 (WHO 2015).

The age-adjusted fracture rate remains stable in North America and Western Europe, but the percentages continue to rise in other parts of the world, such as Asia. Furthermore, even if the incidence rate of hip fractures remains stable, the number of new hip fractures worldwide will increase from 1.7 million in 1990 to 6.3 million in 2050. Assuming that this rate increases by 1% per year, the number of hip fractures worldwide could exceed 8.2 million by 2050 (Kanis et al. 2011; Vijayakumar and Dietrich 2016). The incidence of fractures in the USA is approximately 1.5 million osteoporotic fractures per year. This figure comprises 700,000 vertebral fractures,

300,000 hip fractures, 250,000 wrist fractures and 300,000 other fractures (Kanis et al. 2011). The annual cost of osteoporotic fractures is estimated at 20 billion dollars a year in the United States and 30 billion dollars a year in the European Union (Wu et al. 2018).

In the Caucasian population, around 50% of women and 20% of men over the age of 50 will have at least one fragility fracture in their lifetime (National Osteoporosis Foundation 1998). Prospective studies have shown an increased risk of almost all types of fracture in individuals with low BMD. However, in those who have already had some type of fracture, there is a higher risk of new fractures (Kanis et al. 2011). Given the importance of the disease as highlighted by the epidemiological data, knowledge of its pathophysiology and its possible triggers becomes even more crucial, in order to achieve better control and greater efficacy in the therapeutic response.

2.3 Pathophysiology of Osteoporosis

The loss of bone mass with consequent tissue fragility can currently be classified as Primary OP or Secondary OP (Souza 2010; Fernandes et al. 2015). Primary OP is observed when the origin of the disease is not associated with other previous pathologies or is related to conditions that induce loss of bone mass, such as decreased estrogen levels. Primary OP can be classified into two types: type I or post-menopausal and type II or senile. Type I is mainly related to oestrogen deficiency and is characterized by increased osteoclastic activity, while type II is related to the advanced age of the patients and is characterized by decreased osteoblastic activity (Marques Neto 1997; Guarniero and Oliveira 2004; Cunha et al. 2011). Secondary OP, on the other hand, occurs when it is due to a previous clinical condition such as other diseases, cancer or the intake of medication that induces this loss (Guarniero and Oliveira 2004; Souza 2010; MS 2014).

Female bone formation reaches its peak at around 35 years of age, during which time there is a gradual and increasing loss of bone mass (Fernandes et al. 2015). This process accelerates even more after the menopause, when there is a decrease in estrogen levels with a consequent increase in bone resorption. Estrogen is a hormone with numerous recognized physiological functions, mainly related to growth effects and fertility control (Eriksen et al. 1988). In addition to these functions, Eriksen et al. (1988) added the action of this hormone in human osteoblastic cells carrying estrogen receptors, which was complemented by other studies reporting the presence of these receptors in osteoclasts and osteocytes (Eriksen et al. 1988; Bonnelye et al. 2002; Bord et al. 2003; Faloni and Cerri 2007).

In addition, in a study carried out by Faloni et al. (2007) with rats, the hormone showed the ability to cause apoptosis in osteoclastic cells, the main fact that would relate its decrease to increased bone resorption (Faloni et al. 2007). The main mechanism of action of estrogen and its analogues is in the autocrine and paracrine control of bone homeostasis through the regulation of growth factors and cytokines secreted by osteoblasts or osteoclasts (Faloni and Cerri 2007; Fernandes et al. 2015).

With further research in this area, estrogen has come to be recognized as critical for skeletal homeostasis, with direct and indirect effects on bone cells and their microenvironment (Riancho and Hernández 2012). As a result, the NOF recommendation is that all postmenopausal women should be assessed for BMD levels and the presence of the main associated risk factors (Buttros et al. 2011).

Even with all this care and commitment to discovering the causes of post-menopausal OP, some questions still remain open. In an attempt to fill these gaps, knowledge not only of the action of hormones, which have already been established as participating in homeostasis, but also of the cellular and inorganic constituents and other biological processes involved in the disease are relevant, not only for diagnosis and prognosis, but also for effective therapeutic targets to combat the pathology.

2.4 Bone Tissue

Bone is a dynamic tissue with a high regenerative power made up of cellular components (2%) and bone matrix (98%) with homeostasis guaranteed by various factors such as hormones, vitamins, calcium levels and cytokines, which coordinate the activity of the three cell types that form the tissue: osteoclasts, osteoblasts and osteocytes (Sathy et al., 2015; Lee and Shin 2007; Lewiecki, 2011; Sims and Walsh, 2012; Faienza et al., 2013).

Osteoclasts are large multinucleated phagocytes originating from hematopoietic stem cells that resorb bone through the action of acidic enzymes secreted in the sealing zone resulting from the binding of these cells to the bone surface via integrins with long extracellular domains (Lewiecki, 2011). Its activation is initiated by the action of RANKL secreted by osteoblasts, which binds to the receptor for nuclear factor *kappa* B (RANK) on the osteoclast surface, the whole process being mediated by macrophage colony-stimulating factor (M-CSF). With the activation, proliferation and increased survival of these cells, the whole process of bone resorption begins, which is modulated by the action of the soluble receptor osteoprotegerin (OPG), also secreted by osteoblasts, which competitively inhibits binding between RANKL and its receptor RANK. The expression levels of these cytokines and receptors by bone cells is controlled and mediated by growth factors, hormones, drugs and mainly by the action of other cytokines (Figure 3) (Lewiecki, 2011; Drake et al., 2015).

Resorption is also mediated and compensated for by osteoblasts, cells originating from mesenchymal stem cells, through the secretion of type I collagen and other non-collagenous substances, such as osteopontin, which are responsible for constituting the bone matrix of the tissue removed by the osteoclasts. This process occurs not only in pathology, but also in normal physiological processes (Seeman and Delmas, 2006; Lewiecki, 2011). Stimulation of osteoblast differentiation is induced by activation of the canonical Wnt-p-catenin pathway, through the binding of Wnt signaling proteins to Frizzled receptors associated with the lipoproteins LPR5 and LPR6 and inhibited by the proteins Dickkopf-1 (DKK1) and sclerostin (SOST).

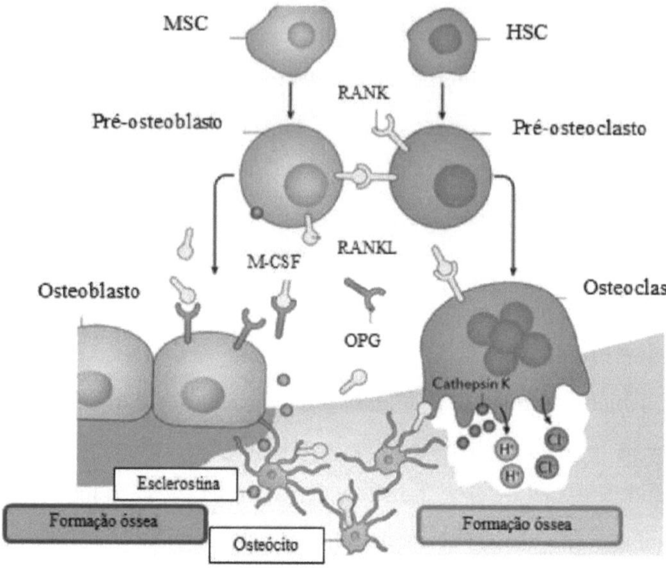

Figure 3: Bone remodeling. The cytokines M-CSF, RANKL, its receptor RANK and OPG acting in the differentiation of bone cell precursors. MSC: mesenchymal stem cell; HSC: hematopoietic stem cell; RANK: receptor for the activator of nuclear factor Kappa B; RANKL: ligand for the receptor for the activator of nuclear factor Kappa B;. MCSF: Colony stimulating growth factor; OPG: Osteoprotegerin; Cathepsin K: Cathepsin K. Source: Adapted from Eastell et al., 2016.

Another anabolic pathway of osteoblastogenesis is that of parathyroid hormone (PTH) and its homologues (PTHrP), controlled by calcium-sensing receptors (CaSR) in the parathyroid glands and which has a direct effect on the differentiation of osteoblasts from osteoprogenitor cells (Lewiecki, 2011; Esbrit et al., 2015). Once their matrix replacement function is complete, osteoblasts go into apoptosis or, if absorbed by the bone matrix, become osteocytes, which regulate the resorption and synthesis of bone tissue (Lewiecki, 2011).

Osteocytes communicate with each other and with surface cells via nerve extensions in bone canaliculi, responding to microcracks and mechanical pressures (*mechanosensation*) and other conditions such as changes in hormone and glucocorticoid levels. This regulation of bone remodeling can promote not only resorption but also tissue replacement (Lewiecki, 2011; Asada et al., 2013). These cells produce both substances that inhibit the Wnt pathway and stimulate osteoclast proliferation, such as sclerostin, and anabolic products that act on osteoblast survival, such as prostaglandins and nitric oxide (Asada et al., 2013). In addition to cellular components, osteocytes also regulate the release of minerals from the bone matrix, in this case preventing their resorption (Pereira, 2009).

The bone matrix has a mineral composition (70%), made up of calcium and phosphate crystals (hydroxyapatite) and an organic composition (30%), made up of type I collagen and other non-collagenous proteins. Glycoproteins and proteoglycans form the amorphous substance on which the hydroxyapatite crystals fit the collagen fibers (Pereira, 2009).

Bone tissue maintains the structure of the human body, protects vital organs, provides movement for muscle contractions, as well as being a reserve of essential ions for life, such as phosphorus and calcium. Dysregulation between bone formation and resorption leads to pathologies resulting from reduced BMD and loss of tissue microarchitecture, such as OP. Although this whole system is multifactorially influenced, the close relationship between the immune system and tissue homeostasis has become prominent, especially after studies involving cytokine genes and bone remodeling, in the emerging Osteoimmunology (Arron and Choi, 2000; Pereira, 2009)
.

2.5 Osteoimmunology

Although several cytokines have already been proven to be related to variations in BMD and the relationship between bone and cells of the immune system is clear due to their place of origin, the term Osteoimmunology was first used by Arron and Choi in 2000, in a review in which the authors emphasized that the relationship between the two goes beyond the immune cells formed in the bone marrow. T lymphocytes and many of their proteins are directly related to bone remodeling, by controlling the action of osteoclasts, osteoblasts and osteocytes (Takayanagi et al., 2000).

The best known pathway in osteoimmunology is RANK/RANKL/OPG, where

RANKL is expressed in T lymphocytes activated by inflammation or a drop in serum estrogen, and binds to RANK in osteoclasts, activating TRAF6, which in turn activates nuclear factor *kappa* B (*NFkB*), Jun-c N-terminal kinase (JNK) and c-Src, responsible for the differentiation, activation and survival of osteoclasts (Figure 4).

At the same time, in a negative *feedback* mechanism, T lymphocytes secrete IFN-Y, which induces the breakdown of the TRAF6 protein or the soluble OPG receptor, secreted by osteoblasts, which bind to RANKL and competitively inhibit its binding to RANK, ensuring the balance of BMD.

However, in pathological states in which this pathway is activated, the resorption caused by the activation of osteoclasts overrides the action of IFN-Y and OPG, causing a decrease in BMD and an increased risk of bone fragility (Arron and Choi, 2000; Takayanagi et al., 2000).

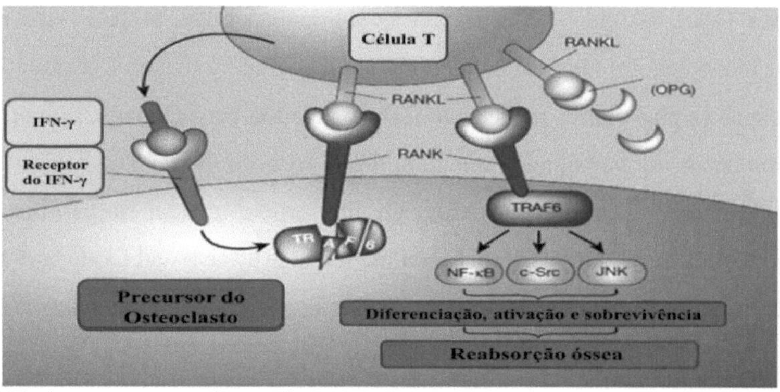

Figure 4: The immune system and bone. The RANKL/RANK/OPG pathway acts on osteoclastogenesis through the interaction of T-cell RANKL with RANK in the osteoclast precursor. OPG and IFN-Y ACT BY inhibiting osteoclastogenesis. Adapted from Arron and Choi, 2000.

Since the Th1 and Th2 responses, through the cytokines IFN-Y and IL-4, are also known to inhibit osteoclastogenesis, the action of other responses, such as Th17, may explain this apparently contradictory picture of the action of T lymphocytes. Th17 cells induce the expression of RANKL in synovial fibroblasts of patients with rheumatoid arthritis, through IL-17 or this cytokine can induce osteoclastogenesis indirectly through the induction of an inflammatory condition with increased expression of TNF-a, IL-6 and IL-1, suggesting that cytokines related to this response may play a fundamental role in preventing bone loss (Takayanagi, 2012) (Figure 5).

Naïve T cells differentiate into Th17 in the presence of IL-23, TGF-p, IL-1 and IL-6, producing mainly IL-17, accompanied by IL-21, IL-22, IL-25 and IL-26 stimulating potent inflammation, while producing IL-10, a cytokine with anti-inflammatory action as a negative *feedback* mechanism (Takayanagi, 2012; Raphael et al., 2014). However, according to some authors, depending on the tissue in question and the stage of the pathology, the IL23/IL17 axis participates in the activation of the Th1 response in a synergistic manner, through the induction of IL-12 and subsequently IFN-Y (Chen et al., 2009; Damsker et al., 2010; Khader and Gopal, 2010).

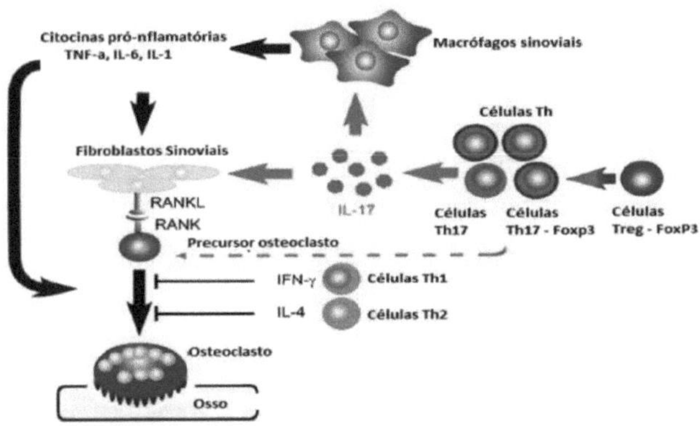

Figure 5: Pro-inflammatory cytokines and osteoclast differentiation. Synovial macrophages activated by IL-17 secrete pro-inflammatory cytokines that stimulate synovial fibroblasts. The cytokines of the Th1 and Th2 responses act to inhibit osteoclastogenesis. Adapted from Guerrini and Takayakagi, 2014.

Due to the recognized action of cytokines in the differentiation, proliferation, activation and survival of bone cells in a direct or indirect way, these proteins represent a potential not only for prognosis, preventing unnecessary treatments and neglecting necessary interventions, but also for therapy, given the various difficulties encountered in current therapeutic guidelines (Takayanagi, 2012; Guerrini and Takayanagi, 2014; Makras et al., 2015).

2.6 Osteoporosis Treatment

The drugs indicated for OP are classified into three groups according to their mechanism of action: anti-catabolic or anti-resorptive, represented by calcium, vitamin D, bisphosphonates, estrogen, selective estrogen receptor modulators, calcitonin and denosumab; anabolic, represented by teriparatide and parathyroid hormone and, finally, those of uncertain action, which probably have both activities, such as strontium ranelate (Riancho and Hernández, 2012).

Calcium administration is indicated for all types of treatment, especially for the risk of vertebral fractures. However, women at risk of heart and kidney disease should give preference to getting calcium from their diet, as some studies suggest that it increases the risk of disease in these organs (Baccaro et al., 2015; Golob and Laya, 2015). The first line of drugs of choice is the bisphosphonate class, represented by alendronate, risendronate, ibandronate and zolendronic acid, which act by preventing osteoclastogenesis through the accumulation of ATP analogues in the cytoplasm of osteoclasts (non-nitrogenous) or by interrupting the mevalonate pathway (nitrogenous), inducing apoptosis and modulating bone resorption (Riancho and Hernández, 2012; MS, 2014). The mechanism of action of estrogen and selective estrogen receptor modulators (SERMS), the latter with fewer side effects, is not yet well defined, but it is believed that they do not act directly to contain osteoclatogenesis, but rather to modulate it by altering cytokine levels with a consequent decrease in bone resorption levels. Calcitonin, on the other hand, acts directly on osteoclasts by binding to membrane receptors and consequently increasing cAMP (Riancho and Hernandez, 2012). Denosumab, approved in 2010 for high-risk OP, is a human anti-RANKL antibody that prevents the protein from binding to its receptor and inhibits the activation and proliferation of osteoclasts, reducing bone resorption (Cummings et al., 2009; Lewiecki, 2011; Riancho and Hernández, 2012).

The anabolic class (Teriparatide or PTH) act differently from antiresorptive drugs, stimulating bone remodeling, increasing not only the amount of bone, but also promoting the restructuring of the tissue's microarchitecture by improving trabecular connectivity (Esbrit et al., 2015; Yoshiki et al., 2016). Its mechanism of action is broad, ranging from direct stimulation of osteoblast proliferation to modulation of the Wnt pathway and stimulation of cytokine expression (Esbrit et al., 2015). In 2016, Eastell et al. compiled results of meta-analyses of the effect of the drugs used on the incidence of hip, vertebral and non-vertebral fractures (Figure 6).

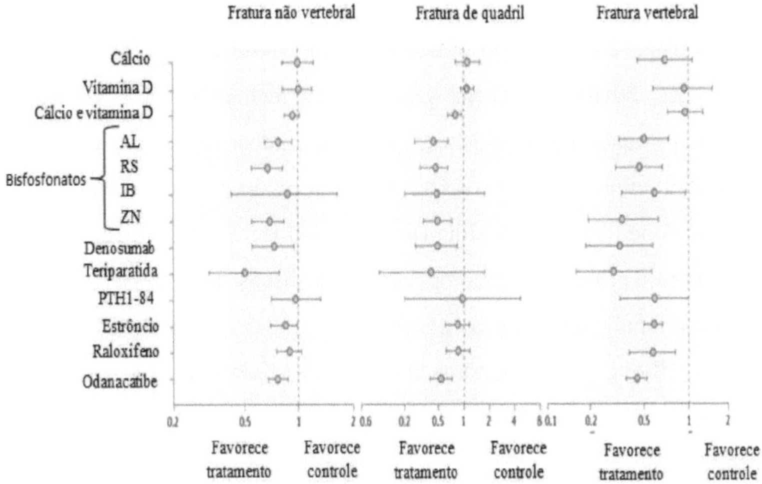

Figure 6 Meta-analyses relating to the incidence of fractures during the various types of therapy used in osteoporosis. AL: Alendronate; RS: Risendronate; IB: Ibandronate; ZN: Zolendronate/Zolendronic acid
Source: Adapted from Eastell et al., 2016.

According to the analysis, among the most commonly used therapies, the use of calcium and vitamin D alone presented a risk of fractures at all the sites analyzed. In the bisphosphonate group, only ibandronate presented a risk for non-vertebral and hip fractures. Teriparatide showed a risk only for the hip, although in the other areas showed greater improvement than all the other treatments. Finally, PTH showed a risk for all the areas studied.

Because it is an emerging public health problem in a relatively recent period, research has turned its attention to the underlying mechanisms of the development of OP, its agents and molecular mechanisms (Eastell et al., 2016). The consolidation of knowledge of these factors represents not only support for an accurate prognosis, but above all for efficient therapy, in view of the morbidity and mortality rates involved in the process. Contradictory results in the action of these components and the scarcity of functional studies involving these elements provide further support for these studies (Blumenfeld et al., 2014).

2.7 Genetics of Osteoporosis

One of the main factors involved in the establishment of OP is heredity, especially for variations in BMD, fracture risk and the area of occurrence of these events (Rivadeneira et al., 2009; Eastell et al. 2016; Karasik et al., 2016; Rocha-Braz and Ferraz-de-Souza, 2016). In recent years, *array* technology and next-generation sequencing have made it possible to expand the number of genes studied and their relationship with clinical aspects of OP, thus suggesting possible pathways acting in the process of bone tissue remodeling (Riancho and Hernández, 2012; Karasik et al., 2016). Recognition of the genetic component in the disease has grown to such an extent that two international consortia were created in the first decade of the 21st century with the aim of identifying genetic variants at risk of the disease: the Genetic *Markers for Osteoporosis Consortium* (GENOMOS) in 2003 and the Genetic Factors for *Osteoporosis Consortium* (GEFOS) in 2008. GEFOS conducted 3 meta-analysis studies based on GWAS in which 36,016 SNPs in more than 150 genes were evaluated. In these studies, a total of more than 240 SNPs located in 23 different genes showed an association with BMD variations, among them the RANKL (TNFSF11), RANK (TNFRSF11A) and OPG (TNFRSF1B) genes. In relation to fractures, a total of 60 SNPs were reported in 5 genes, including RANKL (TNFSF11) and RANK (TNFRSF11A).

This initiative has been essential for advances in knowledge of genetic factors related to the disease, because although rare variants or genes related to autosomal diseases such as SOST and Wnt have been associated with OP, the pathology, being a complex disease, seems to be more influenced by common variants that jointly contribute to susceptibility and phenotypic changes (Riancho and Hernández, 2012; Karasik et al., 2016). BMD, for example, is controlled in 60 to 80% of cases by different genes acting together, according to studies carried out on identical twins (Flicker et al. 1995; Zhou et al., 2014).

In 2014, Liu et al. conducted an *update of* the GWAS carried out on OP, evaluating 19 studies of which 5 were meta-analyses and among these, 2 were carried out by GEFOS. The GWAS identified several genetic polymorphisms in more than 60 genes associated with variations in BMD, particularly in genes for vitamin D receptors (VDR), estrogen receptors (ESR-a and ESR-p) and components of the osteoclast and osteoblastogenesis pathways, such as the RANK/RANKL/OPG, Wnt and MVK pathways (Riancho and Hernández, 2012; Urano and Inoue, 2014; Clark and Duncan, 2015; Karasik et al., 2016). In addition, around 20 genes have been identified as related to fracture risk and more than 100 to OP susceptibility (Liu et al., 2014; Clark and Duncan, 2015; Karasik et al., 2016).

Among these disease-related genes, immune system genes stand out. The IL-21 receptor gene (IL21R) and transforming growth factor beta III receptor (TGFBR3) were associated with changes in BMD in Chinese Caucasian women, while the TNFSF11, TNFSF11A and TNFSF11B genes were related to the osteoclastogenesis process in Chinese and European Caucasian women (Liu et al., 2014; Karasik et al., 2016; Rocha-Braz and Ferraz-de-Souza, 2016). Amid this large number of reported genes, Karasik et al. (2016) suggested a selection of those most closely linked to the disease based on GWAS analyses and functional studies. This selection again mentions the RANK/RANKL/OPG cytokine pathway genes (Figure 7).

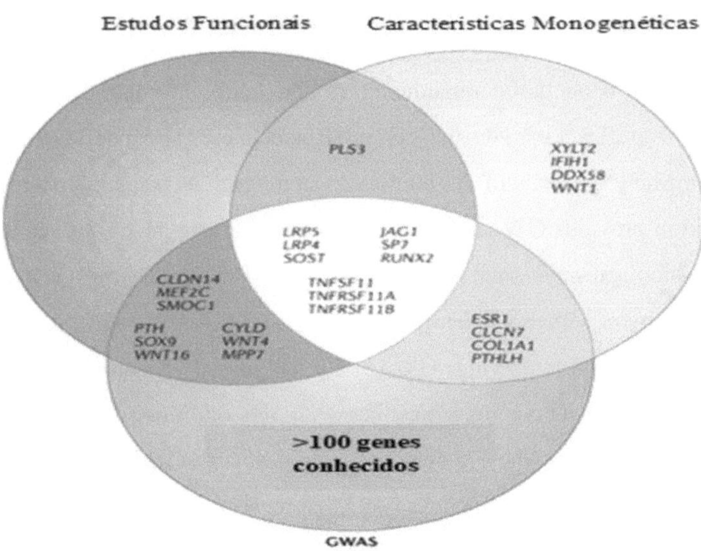

Figure 7 Selection of the main genes associated with Osteoporosis from GWAS and functional studies. Source: Adapted from Karasit et al., 2016.

Despite the high statistical power of these studies, the SNPs identified seem to contribute only a small fraction of changes in phenotypes and susceptibility to disease. Karasik et al. (2016) reported well on the influence of several genes with small effects and the high contribution of environmental factors in complex diseases, such as OP, when compared to monogenic diseases, with the opposite profile (Figure 8).

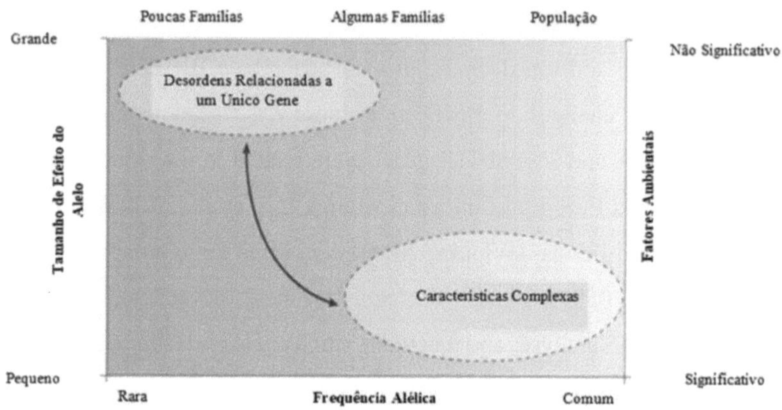

Figure 8: Influence of alleles in monogenic and complex diseases. Source: Adapted from Karasit et al., 2016.

Therefore, these analyses should be used to identify possible pathways acting in the process of bone tissue remodeling as the main information, not limited to statistically significant markers in isolation (Riancho and Hernández, 2012; Karasik et al., 2016). Many elements of the immune system, such as some cytokines, have not been cited in previous GWAS analyses, yet they are recognized for acting in many disease-related pathways, such as MVK, SOST and RANKL/RANK/OPG.

In agreement with this statement, Rivanadeira et al (2009) indicated 13 SNPs that had never been cited in previous GWAS, but had already been reported in common association studies. These data, together with the findings in Osteoimmunology, encourage research into SNPs in cytokine genes associated with these pathways, even if they have not been cited in previous GWAS. Recent studies have described cytokines such as IFN-Y inhibiting the RANK/RANKL/OPG pathway as well as IL-17 inducing this same pathway, both of which indirectly affect bone homeostasis processes (Guerrini and Takayanagi, 2014; Talaat et al., 2015). Functional SNPs in the genes of these cytokines, and their main inducers, IL-23 and IL-12, respectively, have been reported in various diseases, including bone disorders (Wong et al., 2012; Emami et al., 2015; Marques et al., 2015; Zhang et al., 2015; Leng et al., 2016).

Although the importance of these immune system proteins in the establishment of OP is recognized, few studies have been carried out functionally relating these SNPs to susceptibility to post-menopausal OP and its therapeutic response.

2.7.1 SNPs in Cytokine Genes

Cytokines are proteins with essential roles not only in innate and adaptive immunity, but also in influencing and being influenced by various systems and organs such as bone, as well as being associated with the development of various pathologies (Bhushan and Perumal, 2012; Takayanagi, 2012). SNPs can alter both the structure (exon and introns in *splincing* regions) and the levels of these proteins (regulatory regions: 5' and 3' UTR and promoters) and many studies, including the GWAS studies mentioned above, have identified important markers in the susceptibility and manifestations of various multifactorial diseases, such as OP (Richards et al., 2008; Rivadeneira et al., 2009; Mah et al., 2011; Qin et al., 2016). On the other hand, since there is no association with pathologies in another group of studies, many questions are being asked about the role of these SNPs and the interpretations arising from studies with statistically significant associations (Bhushan and Perumal 2012).

In order to derive parameters for the causal relationship between SNPs and diseases, the DACS-DB (*Disease Associated Cytokine SNP Database)* (http://www.iupui.edu/~cytosnp) was created in 2012. This public database contains information on 453 cytokine genes, 63,000 SNPs and 853 diseases associated with these SNPs. Other more general databases that can be consulted for studies are 1000 genomes, Ensembl Genome Browser and the dbSNSP section of the *National Center for Biotechnology Information* (NCBI). These freely accessible sites provide information on the location of polymorphisms, the frequency of the minor allele according to different populations and published scientific studies. The SNP evaluated in this study, IFNG -1616 (G>A, rs2069705), is registered in all of them and although it has been cited in previous studies as being located in the gene's promoter region, it is currently registered as being located intronically on chromosome 12 and has been mentioned in 41 published studies. The SNP has been reported to be associated with mortality outcomes in North American patients (1000 Genomes - http://www.internationalgenome.org).

Lima et al. (2017) (Lima et al. 2017) reported for the first time the relationship between this SNP and biochemical markers of bone remodeling and BMD variations in patients undergoing bisphosphonate treatment. In the study, the IFNG rs2069705 SNP was the one that showed the greatest relationship with therapeutic response, being associated with changes in PTH and BMD values during the 4 years of treatment. These data suggest an important role for this polymorphism in therapeutic decisions in OP.

Computational tools have also been used to measure the effects of SNPs on their corresponding proteins, as well as simulating therapeutic responses. These *in silico* analyses represent a quick and relatively easy way out, compared to the experiments that would need to be carried out to predict the phenotypic effects of the growing number of genes previously associated with diseases (Mah et al., 2011; Singh and Mistry, 2016). These algorithms measure the differences in free energy, solubility and dynamics of the proteins originating from the different SNPs and their wild and mutant variants (Singh and Mistry, 2016). The sequences to be studied can be obtained from available databases and the simulations carried out in software chosen according to the desired analysis. For initial analyses, programs such as I-Mutant and SNP&Go can predict the effect of SNPs on protein stability, while tools such as Swiss-PdbViewer measure the effect of these variants based on the analysis of Force Field Energies and Minimization Energies. Higher energy values correspond to more unstable proteins (Singh and Mistry, 2016; Swiss-PdbViewer tutorial).

All this technological progress in determining potential SNPs related to pathologies and therapeutic responses has allowed potential markers for complex diseases to be identified more quickly and accurately. However, it is important to emphasize the importance of functional validation of these studies to prove the statistical associations of GWAS and computational analyses. *In* this way, we return to the scenarios of expression studies and *in vivo* analyses, using the information generated by new technologies to direct or even ratify the results obtained.

Based on all the information described on the immunogenetic factors involved in the establishment of OP, further study of genes in proteins of the immune system is a promising alternative for a better understanding of this multifactorial disease and its treatment, in view of the role of genetic and immunological components in the balance of bone remodeling.

Chapter 3

Interferon gamma (IFN-Y)

Although technological advances have revealed new genes and numerous polymorphisms associated with OP, the biological function of many cytokines with a recognized role in bone remodeling remain unclear (Karasik et al. 2016; Rocha-Braz and Ferraz-de-Souza 2016). IFN-Y, one of the main examples, has contradictory functions in the bone environment and its actions as anabolic or catabolic agents remain controversial (Takayanagi et al. 2005; Gao et al. 2007; Zhang et al. 2015).

The Interferon family (IFNs) consists of three main types of proteins: Type I (IFN-a, -p, -£, -œ), Type II (IFN-Y) and Type III (IFN-À1 (IL-29), IFN-À2 (IL-28A) and IFN-À3 (IL-28B), which differ in their protein sequences, receptor types, cells where they are produced and gene loci (Pollard et al. 2013; Boxx and Cheng 2016; Krause 2016). The IFNG gene is located on the long arm of chromosome 12 (12q15) and has 4 exons that encode the IFN-y protein secreted mainly by cells of the innate and adaptive immune system in the form of a homodimer (NCBI, 2018). Although this cytokine has mainly been linked to defense against bacterial (Boxx and Cheng 2016) and viral infections (Komatsu et al. 2016), cancer (Kursunel and Esendagli 2016) and autoimmune diseases (Pollard et al. 2013), its role in the bone environment has gained prominence with advances in osteoimmunology (Arron and Choi 2000; Xiong et al. 2016).

IFN-Y is known to influence the survival, activation and differentiation of osteoclasts through the degradation of the Tumor Necrosis Factor Receptor associated with factor 6 (TRAF6) and subsequent blocking of the RANK/RANKL pathway (Arron and Choi 2000). On the other hand, subsequent studies have shown that IFN-y is capable of promoting osteoclastogenesis by inducing cytokines and activating T cells (Gao et al. 2007; Yun gang et al. 2016).

Despite the findings of GWAS on the immune system and bone remodeling, few functional studies with IFN-Y polymorphisms have been carried out to date. Therefore, the main objective of the present study was to evaluate the role of IFN-Y through the study of its polymorphism INFG -1616 (G>A, rs2069705), its impact on serum protein levels and consequent susceptibility to OP. In addition, the effects of the cytokine were evaluated in relation to bone mineralization by calcification of human osteosarcoma cells (SaOs-2).

3.1 Objectives

General

Investigating the influence of the pro-inflammatory cytokine INF-ʏ on susceptibility to primary post-menopausal osteoporosis

Specifics

1. To carry out a genetic association study of the SNP IFNG -1616 (G>A) (rs2069705) in patients with primary postmenopausal OP and healthy individuals from the state of Pernambuco;
2. To comparatively evaluate the expression, in peripheral blood samples, of the INF-y protein in patients with primary postmenopausal OP and healthy individuals;
4. Establish a cellular study model for *in vitro* tests on OP;
5. To evaluate the action of INF-y on the activity of human bone-forming cells.

3.2 Materials and Methods

Study population

A total of 366 women from the Rheumatology Division of the Hospital das Clínicas of the Federal University of Pernambuco underwent BMD measurements by dual-energy X-ray absorptiometry. Women classified as osteoporotic according to World Health Organization criteria were included in the patient group, while those classified as healthy were included in the healthy control group.

The exclusion criteria for the study were the presence of osteopenia, inflammatory diseases, autoimmune diseases, cancer, the use of anti-inflammatory drugs and other rheumatological diseases. This study was approved by the Research Ethics Committee of the Health Sciences Center, Federal University of Pernambuco (CEP / CCS / UFPE n° 513/11) and all participants signed an informed consent form.

DNA isolation and *INFG* genotyping

Genomic DNA was extracted from whole blood using the rapid *mini salting out* method (Bignon and Viña 1995). The genotyping assay was carried out on the ABI 7500 real-time PCR platform (Applied Biosystems, Foster City, CA, USA) with a specific fluorogenic Taqman probe (Taqman Probes, Applied Biosystems, Foster City, CA, USA).

Serum INF-Y **levels of OP patients**

Serum INF-Y levels of OP patients and healthy women were measured with the Bead Array Human (CBA) detection kit, according to the manufacturer's protocol. All the data was obtained on a Becton Dickinson FACScalibur flow cytometer and the analyses were carried out using the BD CBA software (BD Biosciences, USA).

***In vitro* tests**

The SaOs-2 cell line was chosen as the model for the *in vitro* tests. The culture used originated from the public cell bank of the Federal University of Rio de Janeiro (UFRJ) and was kept in a humidified incubator at 37° C with 5% CO_2. For the experiment, the cells were grown in 6-well plates and kept for 15 days in BulletKit™ osteogenic differentiation medium (Lonza - PT4120) to induce calcification.

The confluent cultures were then seeded in 2 x 10^5 cells per pogo and then treated in the presence or absence of recombinant human IFN-Y (Peprotech, Germany). The maximum (1000 U/ml) and minimum (20 U/ml) concentrations were used. After this test, calcification and cell viability were analyzed after 24 hours.

Calcification analysis

Calcified SaOs-2 cells were plated in 24 pogos and incubated at 37 °C and 5% CO2 for 24 h in the presence or absence of IFN-Y (20 U / mL and 1000 U / mL). The cells were stained with alizarin red (pH 4.2, Sigma cat # A5533) for 20 minutes and then washed with deionized water. The calcium deposits were dissolved in a 5% cetylpyridinium chloride solution (Sigma cat # C0732) and the final concentration was measured using a spectrophotometer with wavelength Á = 570 nm (Varioskan Flash -

Thermo Scientific, Rockford, IL, USA).

Analysis of cell viability

Calcified cells were stimulated for 24 h in the presence of IFN-Y and plated in 24 pogos (20 U/mL and 1000 U/mL). The mitochondrial activity test was measured using the methyl tetrazolium bromide (MTT) method according to Yao et al. (2017) (Yao et al., 2017). The MTT reagent solution (Sigma Chemical Co, St Louis, Missouri) (5 mg / mL) was added and the plates remained under the conditions of 37 °C and 5% CO_2, with constant inspection to observe the color change to brown. The absorbance was measured at A = 550 nm with a reference microplate reader (Dynex MRX Revelation Plate Reader, USA).

Statistical analysis

The SNPStats tool (http://bioinfo.iconcologia.net/SNPstats) was used to compare the genotype and allele frequencies of healthy controls and OP patients and to verify the Hardy-Weinberg equilibrium. The t-Student test and analysis of variance (ANOVA) were used to compare mean values between the groups using SPSS software version 18.0. *P-values of* less than 0.05 were considered statistically significant.

3.3 Results

Characteristics of the study population

Of the postmenopausal women selected for the study, 251 (age ± SD) (62.15 ± 12.04) were classified as osteoporotic while 115 (61.22 ± 5.13) were considered healthy controls.

The genotype and allele frequencies of the INFG -1616 polymorphism (G>A, rs2069705) for all the groups studied are shown in Table 1. All the genotype frequencies of patients with OP and the healthy control group were in Hardy-Weinberg equilibrium.

There were significant differences in INFG rs2069705 allele frequencies between OP cases and healthy controls. Compared to the healthy control group, OP patients had a higher G allele frequency, showing this allele to be significantly associated with an increased risk of osteoporosis (OR = 1.45; 95% CI = 1.03 - 2.05; p = 0.03).

Although the group of OP patients showed a higher frequency of INFG rs2069705 GG genotypes than healthy controls (22.4% vs 14.6%), only the dominant model showed a statistically significant association trend (OR = 0.62; 95% CI = 0.38-1.01; p = 0.056) as shown in Table 1.

Serum INF-γ **levels in patients and healthy controls**

In this analysis, the mean positive serum cytokine levels (mean (ranges)) were higher in the healthy controls (2.12 (1.47 - 9.52) pg/mL) than in the group of OP patients (1.03 (0.82 - 12.27) pg/mL).

In the OP group, the INFG rs2069705 GG genotype showed the highest levels (2.60 (2.00 - 12.27) pg/mL) compared to INFG rs2069705 AA (1.80 (1.28 - 2.32) pg/mL) and INFG rs2069705 GA (0.97 (0.61 - 5.81) pg/mL) (Figure 2). The healthy control group showed a reverse pattern compared to the OP patients. The INFG rs2069705 AA genotype showed the highest levels (5.51 (1.86 - 7.36) pg/mL), followed by GA (3.64 (1.47 - 7.69) pg/mL) and INFG rs206970 GG (2.78 (1.56 - 9.42) pg/mL) (Figure 2). Despite all the differences observed between the expression levels in these genotypes, they were not statistically significant. All these results can be seen in Figure 9.

Table 1. Distribution of allele and genotype frequencies of the INFG rs2069705 SNP among patients and healthy women in the control group (p = OP *vs.* control).

Model	Genotypes	OP n (%)	Controls n (%)	OR (95% CI)	p
	AA	65 (26.5)	38 (36.9)	1	
Codominate	GA	125 (51)	50 (48.5)	0.68 (0.41 - 1.15)	0.082
	GG	55 (22.4)	15 (14.6)	0.47 (0.23 - 0.94)	

	AA	65 (26.5)	38 (36.9)	1	
Dominant	GA+GG	180 (73.5)	65 (63.1)	0.62 (0.38 - 1.01)	0.056
	AA+GA	190 (77.5%)	88 (85.4%)	1	
Recessive	GG	55 (22.4%)	15 (14.6%)	0.59 (0.32-1.10)	0.086
	AA+GG	120 (49)	53 (51.5)	1	
Overdominant	GA	125 (51)	50 (48.5)	0.91 (0.57 - 1.44)	0.067
	Alleles				
	A	255 (52)	126 (61)	1	
	G	235 (48)	80 (39)	1.45 (1.03-2.05)	0.03

OP: Osteoporosis; p: p value; OR: odds ratio; CI: confidence interval

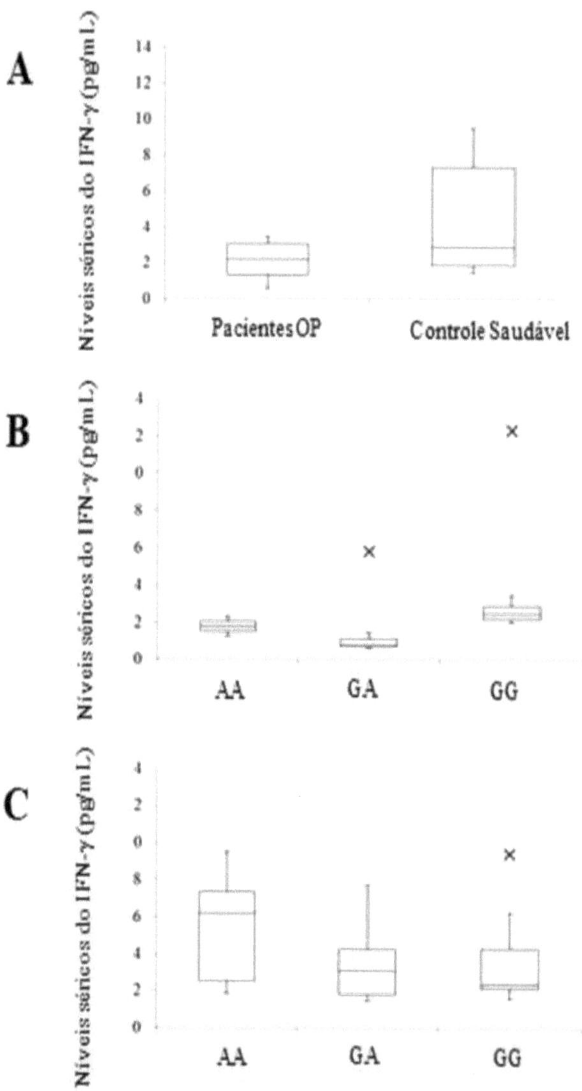

Figure 9 Serum **INF-Y** levels in the healthy control and OP patient groups. A: **INF-Y** levels in the serum of OP patients and healthy control group presented as means (pg/mL). The *boxplots* show the median values (box lines), the 1st quartile (Q1) and the 3rd quartile (Q3) and the deviation bars indicating the minimum and maximum Cq values. The (X) represents the *outlier* values. B: Correlations of serum INF-Y levels with genotype in patients with OP, presented as mean ± SD. C: Correlations of serum **INF-Y** levels with genotype in the healthy control group, presented as mean ± SD.

Analysis of calcification in SaOs-2 cells

INF-y increased calcification *in vitro at* both doses compared to non-stimulated controls. Among the stimulated group, the 20 U/mL concentration of INF-y showed higher levels of calcification than the 1000 U/mL concentration of INF-y (69.14% and 19.15%, higher than the non-stimulated control, respectively) (Figure 10).

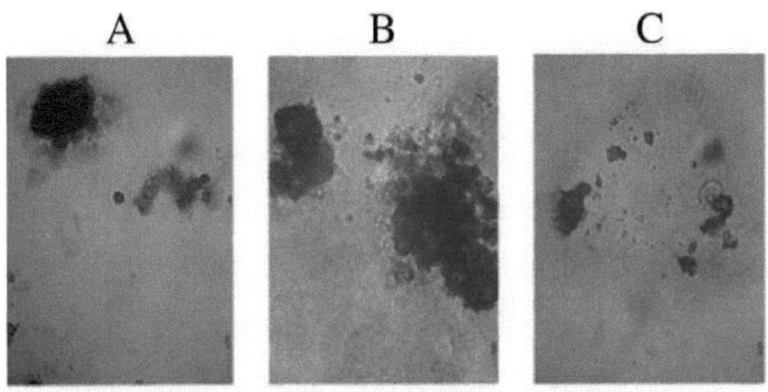

Figure 10 INF-Y increases calcification *in vitro*. SaOs-2 cells were stimulated with INF-Y for 24 h and calcification was quantified by staining with Alizarin Red by spectrophotometry. (A) Unstimulated controls. (B) Cells stimulated with 20 U / mL INF-y. (C) Cells stimulated with 1000 U / mL INF-y.

Viability test

With regard to cell viability, stimuli at concentrations of 20 U / mL and 1000 U / mL increased cell viability compared to the control (7.23% and 13.61%, respectively) (Figure 11). This increase was only significant for the 20 U / mL concentration compared to the unstimulated control group (p = 0.017).

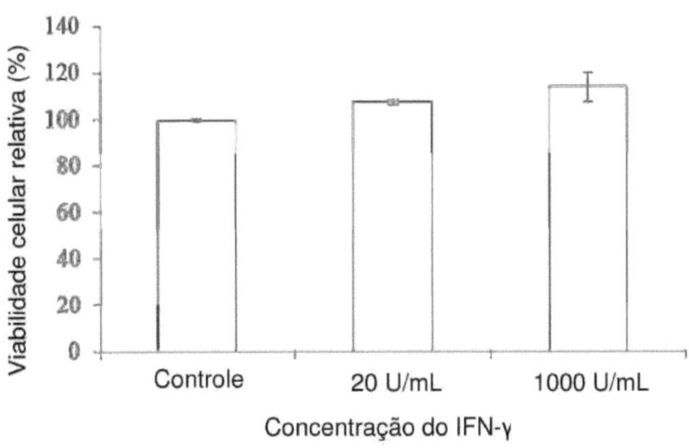

Figure 11 The effects of INF-y on cell viability measured by the 3-(4,5- dimethyl-2-thiazolyl)-2,5-diphenyl-2-H-tetrazolium bromide (MTT) assay after stimulation for 24 h. Data are presented as means ± SD.

3.4 Discussion

To our knowledge, this is the first study to verify the relationship between the IFNG -1616 (G>A) SNP (rs2069705) and primary post-menopausal OP. In a published study, resulting from the thesis that led to this book, the IFNG rs2069705 showed a relationship between its genotypes and clinical characteristics associated with the disease (Lima et al. 2017). The A/A genotype was associated with high levels of PTH, while the G/G genotype was related to high levels of ALP. In addition, there was a relationship between the IFNG rs2069705 G/G genotype and increased BMD in the femoral head and decreased BMD in the total hip, after 3 and 4 years of treatment, respectively. In the present genetic and functional study, subsequent to the previous findings, we also found a relationship between the IFNG gene and its protein with the processes of susceptibility, calcification and cell viability in OP.

Regarding the genetic aspects, we began by confirming the location of the SNP studied, where we found that our polymorphism is in the intronic region and not in the promoter region of the gene, as reported by previous studies. The latest sequencing from 2016, included in the 1000 Genomes Project, located IFNG -1616 (G>A) (rs2069705) in an intronic region, which may give the polymorphism alterations in the protein's stability, but not in its levels. In the analysis of this work, we found that the G variant, associated with susceptibility to OP, seems to confer greater instability to the protein, without altering its levels. Similar to our analysis, previous studies found no changes in IFN-ʏ levels in relation to the different genotypes studied (Erdei et al. 2010; Kim et al. 2010; Huang et al. 2016).

IFN-ʏ has presented contradictory reports on its participation in bone formation and resorption (Gao et al. 2007; Duque et al. 2011). Our work also showed this pleiotropic action of the cytokine in the therapeutic response to bisphosphonates, in the previously published study, with the same INFG rs2069705 G/G genotype associated with different responses in BMD, depending on the site of action and time of disease (Lima et al. 2017). In addition, we found lower concentrations of the cytokine more favorable to calcification and viability in osteoblastic lineage cells than higher concentrations induced in the same assay. These data, combined with previous studies, have enabled us to suggest that the cytokine's action depends on BMD values, the site of action, exposure time to IFN-Y AND THE amount of protein made available. Another important component that we have noticed in previous studies is that the results of the analyses also seem to be related to the types of in *vivo* and *in vitro* models used (Ferrari-Lacraz and Ferrari 2007; Duque et al. 2011). Human bone cells and *in vivo* studies have shown different results when compared to analyses of murine cells (Ferrari-Lacraz and Ferrari 2007; Gao et al. 2007; Duque et al. 2011).

Despite the fact that there is still much to be clarified in relation to the genetic and osteoimmunological factors involved in the establishment of OP, we believe that studies evaluating SNPs, and conducting their functional validations, are of great importance in identifying risk factors and early markers for diagnosis and therapeutic adherence.

In this way, in addition to preventive and personalized treatments based on genetic susceptibility, we could also help to reduce the number of deaths related to around 20% of OP patients who do not adhere to treatment in the first year of therapy (Roush 2011; Iolascon et al. 2013; Drake et al. 2015; Miller 2016). In addition, we believe that these contributions could lead to other prominent research, such as verifying the influence of this polymorphism and the IFN-Y protein on the processes of establishing primary post-menopausal OP.

Conclusions

After the studies carried out in this study, some points regarding Osteoporosis and its genetic and functional factors were clarified, in accordance with the objectives initially proposed.

We concluded that the SNP IFNG -1616 (G>A) (rs2069705) seems to alter the stability of the protein and not its levels. This information is in line with the latest sequencing from 2016 (CRCh38.p7) which proves its position outside the promoter region and its localization in an intronic region. In addition, the IFNG rs2069705 G allele appears to confer greater instability to the protein and is associated with susceptibility to OP in our study population. These data suggest that IFN-Y PLAYS a protective role in OP. In agreement with these results, the protein increased calcification and viability in SaOs-2 cells. However, we believe that this induction is dose-dependent, given that the highest dose (1000 U/mL) used for the cell stimuli showed lower levels of calcification and viability than the minimum value used for the cytokine (20 U/mL).

In addition to answering some of our initial questions, these results encourage functional research in Osteoimmunology, in search of diagnostic and therapeutic markers, as well as bone cell models and optimal concentrations of cytokine stimuli, in order to get closer to understanding the role of these proteins in postmenopausal OP.

Thanks

We would like to thank CAPES (Coordenação de Aperfeigoamento de Pessoal de Nível Superior), CNPq (Conselho Nacional de Desenvolvimento Científico e Tecnológico) and FACEPE (Fundação de Amparo a Ciência e Tecnologia de Pernambuco) for their financial support. We would also like to thank Dr. Joao Ricardo de Oliveira for donating the SaOs-2 cells used in the study.

Bibliographical references

Arron JR and Choi Y (2000) Bone versus immune system. Nature 408:535-536. doi: 10.1038/35046196

Asada N, Katayama Y, Sato M, Minagawa K, Wakahashi K, Kawano H, Kawano Y, Sada A, Ikeda K, Matsui T et al. (2013) Matrix-embedded osteocytes regulate mobilization of hematopoietic stem/progenitor cells. Cell Stem Cell 12:737-747. doi: 10.1016/j.stem.2013.05.001

Aspray TJ (2014) Fragility fracture: recent developments in risk assessment. Ther Adv Musculoskelet Dis 7:17-25. doi: 10.1177/1759720X14564562

Baccaro LF, Conde DM, Costa-Paiva L and Pinto-Neto AM (2015) The epidemiology and management of postmenopausal osteoporosis: A viewpoint from Brazil. Clin Interv Aging 10:583-591. doi: 10.2147/CIA.S54614

Bhushan S and Perumal NB (2012) Disease associated cytokine SNPs database: An annotation and dissemination model. Cytokine 57:107-112. doi: 10.1016/j.cyto.2011.10.009

Bignon JD and Viña MF (1995) 12th IHWC HLA Class II Reference Protocol. Fauchet R Charron D Tech. Handb. 12th Int. Histocompat. Work.

Blumenfeld O, Williams FMK, Valdes A, Hart DJ, Malkin I, Spector TD and Livshits G (2014) Association of interleukin-6 gene polymorphisms with hand osteoarthritis and hand osteoporosis. Cytokine 69:94-101. doi: 10.1016/j.cyto.2014.05.012

Bonnelye E, Kung V, Laplace C, Galson DL and Aubin JE (2002) Estrogen receptor-related receptor a impinges on the estrogen axis in bone: Potential function in osteoporosis. Endocrinology 143:3658-3670. doi: 10.1210/en.2002-220095

Bord S, Ireland DC, Beavan SR and Compston JE (2003) The effects of estrogen on osteoprotegerin, RANKL, and estrogen receptor expression in human osteoblasts. Bone 32:136-141. doi: 10.1016/S8756-3282(02)00953-5

Boxx GM and Cheng G (2016) The Roles of Type i Interferon in Bacterial Infection. Cell Host Microbe 19:760-769. doi: 10.1016/j.chom.2016.05.016

Buttros DDAB, Nahas-Neto J, Nahas EAP, Cangussu LM, Barral ABCR and Kawakami MS (2011) Risk factors for osteoporosis in postmenopausal women from southeast Brazil. Rev Bras Ginecol Obstet 33:295-302. doi: S0100-72032011000600006 [pii]

Chen B, Zeng Z, Xu L, Wu X, Yu J, Xue L, Hao Y, Wang Y, Sung J, Chen M et al. (2009) IL23R +2199A/C polymorphism is associated with decreased risk of certain subtypes of gastric cancer in Chinese: A case-control study. Cancer Epidemiol 35(2):165-169. doi: 10.1016/j.canep.2010.08.006

Chesnut CH, McClung MR, Ensrud KE, Bell NH, Genant HK, Harris ST, Singer FR, Stock JL, Yood RA, Delmas PD et al. (1995) Alendronate

treatment of the postmenopausal osteoporotic woman: Effect of multiple dosages on bone mass and bone remodeling. Am J Med 99:144-152. doi: 10.1016/S0002-9343(99)80134-X
Clark GR and Duncan EL (2015) The genetics of osteoporosis. Br Med Bull 113:73-81. doi: 10.1093/bmb/ldu042
Cummings SR, San Martin J, McClung MR, Siris ES, Eastell R, Reid IR, Delmas P, Zoog HB, Austin M, Wang A et al. (2009) Denosumab for prevention of fractures in postmenopausal women with osteoporosis. N Engl J Med 361:756-765. doi: 10.1056/NEJMoa0809493
Cunha, EP; Steiner, ML; Strufaldi, R; Fernandes, C; Laurindo, IMM; Pereira, RMR; Simões R (2011) Diretrizes Clínicas na Saúde Suplementar - Osteoporose: tratamento. Diretrizes Clínicas Saúde Supl Fed Bras das Assocs Ginecol e Obs Soc Bras Reumatol 19p.
Damsker JM, Hansen AM and Caspi RR (2010) Th1 and Th17 cells. Ann N Y Acad Sci 211-221. doi: 10.1111/j.1749-6632.2009.05133.x.Th1
Drake M, Clarke B and Lewiecki E (2015) The Pathophysiology and Treatment of Osteoporosis. Clin Ther 37:1837-1850
Duque G, Huang DC, Dion N, Macoritto M, Rivas D, Li W, Yang XF, Li J, Lian J, Marino FT et al. (2011) Interferon-g Plays a Role in Bone Formation In Vivo and Rescues Osteoporosis in Ovariectomized Mice. 26:1472-1483. doi: 10.1002/jbmr.350
Eastell R, O'Neill TW, Hofbauer LC, Langdahl B, Reid IR, Gold DT and Cummings SR (2016) Postmenopausal osteoporosis. Nat Rev Dis Prim 2:16069. doi: 10.1038/nrdp.2016.69
Emami S, Ziaee V, Rezaei A, Sadr M, Maddah M, Amirzargar AA and Rezaei N (2015) IL23R gene polymorphism with juvenile idiopathic arthritis and its association with serum IL-17A. Int J Rheum Dis 19(11):1189-1196. doi: 10.1111/1756-185X.12674
Erdei E, Kang H, Meisner A, White K, Pickett G, Baca C, Royce M and Berwick M (2010) Polymorphisms in cytokine genes and serum cytokine levels among New Mexican women with and without breast cancer. Cytokine 51:18-24. doi: 10.1016/j.cyto.2010.03.014
Eriksen E, Colvard D, Berg N, Graham M, Mann K, Spelsberg T and Riggs B (1988) Evidence of estrogen receptors in normal human osteoblast-like cells.
Science 241:8486. doi: 10.1126/science.3388021
Esbrit P, Herrera S, Portal-Núñez S, Nogués X and Díez-Pérez A (2015) Parathyroid Hormone-Related Protein Analogs as Osteoporosis Therapies. Calcif Tissue Int 98(4):359-369 doi: 10.1007/s00223-015-0050-1
Faienza MF, Ventura A, Marzano F and Cavallo L (2013) Review Article Postmenopausal Osteoporosis: The Role of Immune System Cells. 2013: Clin Dev Immunol 2013: 575936
Faloni ADS and Cerri P (2007) Cellular and molecular mechanisms of estrogen in bone resorption. Rev Odontol da UNESP 36:181-188
Faloni APS, Sasso-Cerri E, Katchburian E and Cerri PS (2007) Decrease in

the number and apoptosis of alveolar bone osteoclasts in estrogen-treated rats. J Periodontal Res 42:193-201. doi: 10.1111/j.1600-0765.2006.00932.x
Fernandes TRL, Oliveira JB, Lorencete TV and Amadei JL (2015) Factors associated with osteoporosis in postmenopausal women. Semin Ciencias Biológicas e da Saúde 36:93-106. doi: 10.5433/1679-0367.201v36n1p93
Ferrari-Lacraz S and Ferrari S (2007) Is IFN-γ involved in bone loss or protection? nothing is simple with cytokines. BoneKEy-Osteovision 4:83-87. doi: 10.1138/20060251
Flicker L, Hopper JL, Rodgers L, Kaymakci B, Green RM and Wark JD (1995) Bone density determinants in elderly women: A twin study. J Bone Miner Res 10:1607-1613. doi: 10.1002/jbmr.5650101102
Marques Neto JF (1997) Osteoporosis: Epidemiology. Acta Fisiátrica (USP) 4:119-122
Gao Y, Grassi F, Ryan MR, Terauchi M, Page K, Yang X, Weitzmann MN and Pacifici R (2007) IFN-gamma stimulates osteoclast formation and bone loss in vivo via antigen-driven T cell activation. J Clin Invest 117:122-132. doi: 10.1172/JCI30074
Golob AL and Laya MB (2015) Osteoporosis: screening, prevention, and management. Med Clin North Am 99:587-606. doi: 10.1016/j.mcna.2015.01.010
Greenblatt MB, Tsai JN and Wein MN (2016) Bone Turnover Markers in the Diagnosis and Monitoring of Metabolic Bone Disease. Clin Chem. doi: 10.1373/clinchem.2016.259085
Guarniero R and Oliveira S (2004) Osteoporosis: an update in diagnosis and basic treatment principles. 39:477-485
Guerrini MM and Takayanagi H (2014) The immune system, bone and RANKL. Arch Biochem Biophys 561:118-123. doi: 10.1016/j.abb.2014.06.003
Huang Y-H, Hsu Y-W, Lu H-F, Sung H, Wong -Ching, Yu H-R, Kuo H-C, Huang F-C, Chang W-C and Kuo H-C (2016) Interferon-gamma Genetic Polymorphism and Expression in Kawasaki Disease. 95(17):e3501. doi: 10.1097/MD.0000000000003501
Iolascon G, Napolano R, Gioia M, Moretti A, Riccio I and Gimigliano F (2013) The contribution of cortical and trabecular tissues to bone strength: Insights from denosumab studies. Clin Cases Miner Bone Metab 10:47-51. doi: 10.11138/ccmbm/2013.10.1.047
Kanis JA (2008) Assessment of Osteoporosis at the Primary Health Care Level. WHO Collaborating Centre for Metabolic Bone Diseases, University of Sheffield Medical School, Brussels.
Kanis JA, Cooper C, Hiligsmann M, Rabenda V, Reginster JY and Rizzoli R (2011) Partial adherence: A new perspective on health economic assessment in osteoporosis. Osteoporos Int 22:2565-2573. doi: 10.1007/s00198-011- 1668-0
Karasik D, Rivadeneira F and Johnson ML (2016) The genetics of bone mass and susceptibility to bone diseases. Nat Rev Rheumatol 12:323-334.

doi: 10.1038/nrrheum.2016.48
Khader S and Gopal R (2010) IL-17 in protective immunity to intracellular pathogens. Virulence 1:423-427. doi: 10.4161/viru.1.5.12862
Kim K, Cho S-K, Sestak A, Namjou B, Kang C and Bae S-C (2010) Interferongamma gene polymorphisms associated with susceptibility to systemic lupus erythematosus. Ann Rheum Dis 69:1247-1250. doi: 10.1136/ard.2009.117572
Komatsu T, Nagata K and Wodrich H (2016) The role of nuclear antiviral factors against invading DNA viruses: The immediate fate of incoming viral genomes. Viruses 8(10). pii: E290. doi: 10.3390/v8100290
Krause CD (2016) Intron loss in interferon genes follows a distinct set of stages, and may confer an evolutionary advantage. Cytokine 83:193-205. doi: 10.1016/j.cyto.2016.04.018
Kursunel MA and Esendagli G (2016) The untold story of IFN-γ in cancer biology. Cytokine Growth Factor Rev 31:73-81. doi: 10.1016/j.cytogfr.2016.07.005
Lee SH and Shin H (2007) Matrices and scaffolds for delivery of bioactive molecules in bone and cartilage tissue engineering. Adv Drug Deliv Rev 59:339-359. doi: 10.1016/j.addr.2007.03.016
Leng RX, Pan HF, Liu J, Yang XK, Zhang C, Tao SS, Wang DG, Li XM, Li XP, Yang W et al. (2016) Evidence for genetic association of TBX21 and IFNG with systemic lupus erythematosus in a Chinese Han population. Sci Rep 6:22081. doi: 10.1038/srep22081
Lewiecki EM (2011) New targets for intervention in the treatment of postmenopausal osteoporosis. Nat Rev Rheumatol 7:631-638. doi: 10.1038/nrrheum.2011.130
Lima CAD, Javorski NR, Souza APO, Barbosa AD, Valença APMC, Crovella S, Souza PRE, De Azevedo Silva J and Sandrin-Garcia P (2017) Polymorphisms in key bone modulator cytokines genes influence bisphosphonates therapy in postmenopausal women. Inflammopharmacology 25:191-201. doi: 10.1007/s10787-017-0322-7
Liu Y-J, Zhang L, Papasian CJ and Deng H-W (2014) Genome-wide Association Studies for Osteoporosis: A 2013 Update. J Bone Metab 21:99-116. doi: 10.11005/jbm.2014.21.2.99
Mah JTL, Low ESH and Lee E (2011) In silico SNP analysis and bioinformatics tools: A review of the state of the art to aid drug discovery. Drug Discov Today 16:800-809. doi: 10.1016/j.drudis.2011.07.005
Makras P, Delaroudis S and Anastasilakis AD (2015) Novel therapies for osteoporosis. Metabolism 64:1199-1214. doi: 10.1016/j.metabol.2015.07.011
Marques A, Ferreira RJO, Santos E, Loza E, Carmona L and da Silva JAP (2015) The accuracy of osteoporotic fracture risk prediction tools: a systematic review and meta-analysis. Ann Rheum Dis annrheumdis-2015-207907. doi: 10.1136/annrheumdis-2015-207907
Miller PD (2016) Underdiagnoses and undertreatment of osteoporosis: The

battle to be won. J Clin Endocrinol Metab 101:852-859. doi: 10.1210/jc.2015- 3156
National Osteoporosis Foundation (1998) Osteoporosis International. Osteoporos Rev Evid Prev Diagnosis Treat Cost-Effective Anal 4:S7-S80.
Pereira FB (2009) Analysis of the correlation between muscle strength and bone mineral density in men aged 60 and over living in sâo sebastiâo - DF. Master's Degree in Health Sciences, University of Brasilia, Federal District
Pollard KM, Cauvi DM, Toomey CB, Morris K V and Kono DH (2013) Interferon- Y and systemic autoimmunity. Discov Med 16:123-131. doi: 10.1016/j.biotechadv.2011.08.021.Secreted
Qin L, Liu Y, Wang Y, Wu G, Chen J, Ye W, Yang J and Huang Q (2016) Computational characterization of osteoporosis associated SNPs and genes identified by genome-wide association studies. PLoS One 11(3):e0150070. doi: 10.1371/journal.pone.0150070
Raphael I, Nalawade S, Eagar TN and Forsthuber TG (2014) T cell subsets and their signature cytokines in autoimmune and inflammatory diseases. Cytokine 74(1):5-17. doi: 10.1016/j.cyto.2014.09.011
Riancho J and Hernández JL (2012) Pharmacogenomics of osteoporosis: a pathway approach. Pharmacogenomics 13:815-829. doi: 10.2217/pgs.12.50
Richards J, Rivadeneira F, Inouye M, Pastinen T, Soranzo N, Wilson S, Andrew T, Falchi M, Gwilliam R, Ahmadi K et al. (2008) Bone mineral density, osteoporosis, and osteoporotic fractures: a genome-wide association study. Lancet 371:1505-1512. doi: 10.1016/S0140-6736(08)60599-1
Rivadeneira F, Styrkársdottir U, Estrada K, Halldórsson B V, Aulchenko YS, Cupples LA, Deloukas P, Demissie S, Grundberg E, Hofman A et al. (2009) Twenty bone-mineral-density loci identified by large-scale meta-analysis of genome-wide association studies. Nat Genet 41(11):1199-206. doi: 10.1038/ng.446
Rocha-Braz MGM and Ferraz-de-Souza B (2016) Genetics of osteoporosis: searching for candidate genes for bone fragility. Arch Endocrinol Metab 60:391-401. doi: 10.1590/2359-3997000000178
Roush K (2011) Prevention and treatment of osteoporosis in postmenopausal women: a review. Ajn 111:26-35
Sathy BN, Watson BM, Kinard LA and Spicer PP Bone tissue engineering with multi-layered scaffolds - Part II: Combining vascularization with bone formation in critical sized bone defect. Tissue Eng Part A 21(19-20):2495-2503.
Seeman E and Delmas PD (2006) Bone quality--the material and structural basis of bone strength and fragility. N Engl J Med 354:2250-2261. doi: 10.1056/NEJMra053077
Seibel MJ (2005) Biochemical markers of bone turnover: Part I: biochemistry and variability. Clin Biochem Rev 26:97-122.
Sims N and Walsh NC (2012) Intercellular cross-talk among bone cells: New factors and pathways. Curr Osteoporos Rep 10:109-117. doi:

10.1007/s11914-012-0096-1
Singh PK and Mistry KN (2016) A computational approach to determine susceptibility to cancer by evaluating the deleterious effect of nsSNP in XRCC1 gene on binding interaction of XRCC1 protein with ligase III. Gene 576:141-149. doi: 10.1016/j.gene.2015.09.084
Souza MPG (2010) Diagnosis and treatment of osteoporosis. Rev Bras Ortop 45:220-229. doi: 10.1590/S0102-36162010000300002
Takayanagi H (2012) New developments in osteoimmunology. Nat Rev Rheumatol 8:684-689. doi: 10.1038/nrrheum.2012.167
Takayanagi H, Ogasawara K, Hida S, Chiba T, Murata S, Sato K, Takaoka A, Yokochi T, Oda H, Tanaka K et al. (2000) T-cell-mediated regulation of osteoclastogenesis by signaling cross-talk between RANKL and IFN-gamma. Nature 408:600-605. doi: 10.1038/35046102
Takayanagi H, Sato K, Takaoka A and Taniguchi T (2005) Interplay between interferon and other cytokine systems in bone metabolism. Immunol Rev 208:181-193. doi: 10.1111/j.0105-2896.2005.00337.x
Talaat RM, Sidek A, Mosalem A and Kholief A (2015) Effect of bisphosphonates treatment on cytokine imbalance between TH17 and Treg in osteoporosis. Inflammopharmacology 23:119-125. doi: 10.1007/s10787-015-0233-4
Urano T and Inoue S (2014) Genetics of osteoporosis. Biochem Biophys Res Commun 452:287-293. doi: 10.1016/j.bbrc.2014.07.141
Vijayakumar R and Dietrich B (2016) Review article Osteoporosis: An underrecognized public health problem. J Local Glob Heal Sci. 1(2):71-77 doi: dx.doi.org/10.5339/ jlghs.2016.2
WHO (2015) World Health Organization. World report on ageing and health.260p.
Wong R, Wei J, Huang C, Lee H, Chiou S, Lin S, Cai Y, Hung P, Wang M and Yang S (2012) Association of IL-12B genetic polymorphism with the susceptibility and disease severity of ankylosing spondylitis. J Rheumatol 39:135-140. doi: doi: 10.3899/jrheum.110613
Wu C-H, Kao I-J, Hung W-C, Lin S-C, Liu H-C, Hsieh M-H, Bagga S, Achra M, Cheng T-T and Yang R-S (2018) Economic impact and cost-effectiveness of fracture liaison services: a systematic review of the literature. Osteoporos Int [Epub ahead of print]. doi: 10.1007/s00198-018-4411-2
Xiong Q, Zhang L, Ge W and Tang P (2016) The roles of interferons in osteoclasts and osteoclastogenesis. Jt Bone Spine 83:276-281. doi: 10.1016/j.jbspin.2015.07.010
Yoshiki F, Nishikawa A, Taketsuna M, Kajimoto K and Enomoto H (2016) Efficacy and safety of teriparatide in bisphosphonate-pretreated and treatment-naive patients with osteoporosis at high risk of fracture: Post hoc analysis of a prospective observational study. J Orthop Sci 22(2):330-338. doi: 10.1016/j.jos.2016.11.022
Yun gang L, Kang L and Wu G (2016) Fluorosis increases the risk of

postmenopausal osteoporosis by stimulating interferon γ- Biochem Biophys Res Commun 479:372-379. doi: 10.1016/j.bbrc.2016.09.083

Zaia A (2015) Fractal lacunarity of trabecular bone and magnetic resonance imaging: New perspectives for osteoporotic fracture risk assessment. World J Orthop 6(2):221-235. doi: 10.5312/wjo.v6.i2.221

Zhang J, Fu Q, Ren Z, Wang Y, Wang C, Shen T, Wang G and Wu L (2015) Changes of serum cytokines-related Th1/Th2/Th17 concentration in patients with postmenopausal osteoporosis. Gynecol Endocrinol 31:183-190. doi: 10.3109/09513590.2014.975683

Zhou P, Liu H, Liao E, Zhang Z, Chen de C, Liu J, Wu W, Xing X, Xia W, Xu L et al. (2014) LRP5 polymorphisms and response to alendronate treatment in Chinese postmenopausal women with osteoporosis. Pharmacogenomics 15:821-831. doi: 10.2217/pgs.14.12

Zhou P, Xu X, Zhang Z, Liao E, Chen D, Liu J, Wu W, Jiang Y, Wang O, Xia W et al. (2015) SOST polymorphisms and response to alendronate treatment in postmenopausal Chinese women with osteoporosis. Pharmacogenomics 16:1077-1088

Zupan J, Jeras M and Marc J (2013) Osteoimmunology and the influence of pro- inflammatory cytokines on osteoclasts. Biochem Medica 23:43-63. doi: 10.11613/BM.2013.007

MIX
Papier aus verantwortungsvollen Quellen
Paper from responsible sources
FSC® C105338

Printed by Books on Demand GmbH, Norderstedt / Germany